TRAVELS OF A PHOTOGRAPHER IN CHINA
1933–1946

TRAVELS OF
A PHOTOGRAPHER
IN CHINA
1933-1946

Hedda Morrison

HONG KONG OXFORD NEW YORK
OXFORD UNIVERSITY PRESS
1987

Oxford University Press

Oxford   New York   Toronto
Petaling Jaya   Singapore   Hong Kong   Tokyo
Delhi   Bombay   Calcutta   Madras   Karachi
Nairobi   Dar es Salaam   Cape Town
Melbourne   Auckland

and associated companies in
Beirut   Berlin   Ibadan   Nicosia

First published 1987
First published in the
United States by Oxford
University Press, Inc., New York

British Library Cataloguing in Publication Data

Morrison, Hedda
Travels of a photographer in China, 1933–1946.
1. China—Description and travel—1901–1948
I. Title
915.1'0442   DS710
ISBN 0-19-584098-4
ISBN 0-19-584177-8 Pbk

Library of Congress Cataloging-in-Publication Data

Morrison, Hedda.
Travels of a photographer in China, 1933–1946.

Bibliography: p.
1. China—Description and travel—1901–1948—Views.
I. Title.
DS710.M495   1987   915.1'0442   87-7648
ISBN 0-19-584098-4
ISBN 0-19-584177-8 (pbk.)

Printed in Hong Kong by Golden Cup Printing Co. Ltd.
Published by Oxford University Press, Warwick House, Hong Kong

Preface

THE kind reception that has been accorded my book on old Peking (*A Photographer in Old Peking*, 1985) has prompted me to put together a collection of some additional photographs taken outside the confines of the city during my stay in China between 1933 and 1946. Before the Japanese invasion in 1937 travel could easily be undertaken. It was possible to travel simply and at low cost: you could go where you pleased and as you pleased. Even so I was not able to travel nearly as much as I would have liked. I had a six-day working week and few holidays, and no experience of China to guide me when I first arrived. After 1937 movement was greatly restricted by the Japanese invasion and occupation. I hope, nevertheless, that my photographs may prove of interest to those who did not have the opportunity to see something of the China that existed before the great changes that have occurred since 1949.

Relief sculpture in the rear chamber of Cave 8.

Relief sculpture in the rear chamber of Cave 8.

Apsaras and donors on the north wall of Cave 44.

Despite their battered condition, what had survived of
the caves and their sculpture conveyed a vivid impression
of their original magnificence.

Many of the heads were restorations, the originals
scattered in Western museums and collections.

Rows of small seated Buddhas in individual niches.

Interior of a dilapidated temple.

Pagodas often survived better than other buildings.

Grotesque figure supporting the base of a pagoda.

Printing block lying in a temple courtyard.

Jehol 1934 and 1935

I made use of my first holiday in the summer of 1934 to visit Jehol (Ch'eng Te), the old summer seat of government and imperial residence 230 kilometres north-east of Peking. Ch'eng Te is the name of the Chinese town just south of the palaces and temples. The complex of palaces and temples came to be known by foreigners as Jehol, a corrupt version of the name of the river Je Ho (literally Hot River) on the banks of which the town, palaces and temples are all situated. The imperial edifices were built by Emperor K'ang Hsi (reigned 1662–1722) and enlarged and added to by his successors. They were for many years regularly occupied during the summer months by the court anxious to escape from the burning heat of Peking.

In 1820 Emperor Chia Ch'ing was struck by lightning in Jehol. Not unreasonably this was considered to be an inauspicious omen, and Jehol was not used again for forty years until 1860 when Emperor Hsien Feng took refuge there as Peking was occupied by British and French troops. His death in Jehol less than a year later confirmed the worst suspicions of the court as to the locality's malign influences and general insalubrity and the court never returned to it. When Peking was again occupied by foreign troops in 1900 and the Empress Dowager had to flee the capital, she chose to go to Sian. In the 1930s many of the buildings in Jehol were still standing and it remained a most impressive monument to the power and wealth of imperial China.

The palaces were situated to the west of Je Ho in a great walled pleasure ground, the walls some 10 kilometres in length. Both north and south of this park and again on the east bank of Je Ho are a number of fine temples and monasteries. Two of the latter are of Tibetan design, one of them being a copy of the Potala in Lhasa. Indeed, Jehol was not only a summer resort for the emperor but also played an important part in the empire's public relations, as it were, with often troublesome border peoples, the nomads of the north-west and the Tibetans. These peoples were followers of the Lamaist form of Buddhism, as were the Manchus themselves. It made sound political sense to have an impressive centre of Lamaist Buddhism and of imperial power on the Mongolian marches of the empire.

Religion and politics apart, Jehol is a very beautiful place on a great bend of the river ringed by mountains. In imperial days it was well wooded, but by the thirties most of the trees had long since disappeared. Regrettable though this was, their absence enabled one to appreciate more readily the grand layout of Jehol's palaces and temples.

I set off for Jehol on the back of a truck loaded with bags of flour. The truck people obligingly picked me up at the Wagons-Lits Hotel where I occupied a small attic room. I think the manner of my departure occasioned a little surprise on the part of some fellow hotel guests who watched me leave. In imperial days the road from Peking to Jehol was the finest road in China. This unfortunately was no longer the case and we encountered problems on the way. There had been recent heavy rain and many streams were in flood. At the ford across one stream the engine died when we had only partially negotiated the way over, and the passengers had to complete the crossing on foot. I was helped across by a burly Korean fellow passenger whose command of English was limited to the words 'Never mind'. We had to spend a night at an inn while we waited for the truck to complete the crossing.

In Jehol I met the local Catholic missionary, a kind and well-informed Belgian of liberal outlook, who insisted that I stay at the mission. He told me that as a general rule Catholic missionaries were not supposed to entertain women but that this ruling might easily be overlooked. He asked one of the Chinese nuns at the mission to accompany me on my tours of the ruins and remaining buildings. She smoked a pipe and the only indication of her calling was a rosary. I could not help noticing that it was her custom to make polite obeisance before images of the Buddha.

In 1934 Jehol was already under Japanese occupation. The local Japanese commander invited me to dinner, but my kind Belgian host did not feel that it would be entirely proper for me to accept the invitation. Instead, he asked several of the Japanese officers to dine at the mission.

I spent several days exploring Jehol and paid another visit in the following year after my return from Hua Shan. This enabled me to take a further set of photographs with a large-format camera that was more suited to architectural studies than the Rolleiflex which I had had with me on the first occasion.

Jehol has been described by the famous explorer Sven Hedin in a book published in 1933. He came on his last expedition to Asia while I was still working at Hartungs, the photographic studio. After the expedition's return to Peking, Mr Lu, the Hartungs cine photographer, and I accompanied it on an outing to the Ming Tombs. It was a hilarious occasion. Hedin needed some footage from the Ming Tombs which, in their then deforested state, looked very much as if they were in Central Asia. The outing involved the setting up of a real explorer's tent while Mr Lu photographed Hedin with one of the cumbersome 35-mm cine cameras in use at that time.

Bogged down en route to Jehol.

P'ai-lou on the way to the Small Potala.

The Small Potala, which is a copy of the Potala in Lhasa, designed
to make visiting Tibetans feel at home and to demonstrate
the Ch'ing government's appreciation of Tibetan culture.

The Small Potala.

Figure of the Big Buddha in P'u Ning Ssu, portrayed with a third eye in the forehead.

P'u Ning Ssu, popularly known as Ta Fo Ssu, the Big Buddha Temple.

View over Hsü Mi Fu Shou Chih Miao with the pagoda in the background.

Pagoda behind Hsü Mi Fu Shou Chih Miao.

Hsü Mi Fu Shou Chih Miao, a building partly in Tibetan style.

Roof of a pavilion within Hsü Mi Fu Shou Chih Miao.

The golden-roofed pavilion of Hsü Mi Fu Shou Chih Miao was surmounted by eight dragons.

Ferocious figure from the Lamaist pantheon.

Ghostly Lamaist figures on a wall painting.

Statue of a lama in Hsü Mi Fu Shou Chih Miao.

Caryatidic figures in Hsü Mi Fu Shou Chih Miao.

Buddhist wall painting.

P'u Lo Ssu, a temple in Chinese style.

Hua Shan 1935

THE suggestion that I visit the sacred Taoist mountain of Hua Shan came from Henri Vetch, the proprietor of the French Book Store in Peking. Vetch, who was extremely interested in Taoism, had never been to Hua Shan himself but knew of its beauty. Hua Shan is a most spectacular outlier of the range known as Tsin Ling in eastern Shensi province. It lies to the east of the ancient city of Sian and overlooks the narrow plain through which flows the Yellow River, bounded by the Tsin Ling range and to the north the mountains of Shansi. Hua Shan consists of a ring of precipitous peaks rising to 2,500 metres in height, enclosed in the centre of which are small patches of relatively flat land. The peaks are separated from the main Tsin Ling range by awe-inspiring chasms.

The mountain has been held sacred in China from very early times, and religious observances are said to have been held here from as far back as 1760 BC. It was therefore natural that it should become a Taoist centre when that religion was founded about 2,000 years ago. The importance of Hua Shan lay partly in its proximity to the early centres of Chinese civilization, and partly in its being a supreme example of the type of landscape so appreciated by Chinese artists. There was, however, much more to Hua Shan than its spectacular mountain forms. The solitary, cloud-capped peaks were endowed with mysticism in the Chinese mind, and this, together with the allied philosophical concept of man as an insignificant creature in cosmic nature, found expression in the works of the landscape painters of the Sung dynasty.

I visited Hua Shan in the summer of 1935. It was an easy place to reach because the train to Sian stopped conveniently at a little place called Hua Yin, which means 'under the shadow of Hua Shan'. The Hua Shan massif did indeed dominate Hua Yin. I took a rickshaw from the station to the fine Jade Spring Temple picturesquely situated at the start of the ascent to the mountain. The temple extended hospitality to the many pilgrims who visited Hua Shan and I spent a comfortable night there after an appetizing vegetarian meal.

The next day I hired a porter to carry my rucksack and set off for the mountain. The way lay up a narrow defile between rock faces hundreds of metres high. At convenient intervals there were small temples providing hot tea for visitors. I noticed that my porter, apart from taking breaks for tea, stopped several times for a quiet smoke, not of tobacco, but of opium. At midday we reached the end of the defile and from there the real climb began. It was very steep. In places the track led up almost perpendicular rock faces in which steps had been hewn and iron chains of uncertain reliability set in the rock to provide hand-holds. Some of these sheer stretches were for one-way traffic only, and when we reached them we would call out so as to ensure that we did not meet some descending pilgrim half way. I did indeed meet a number of other visitors as pilgrimage to the mountain continued throughout the year.

Eventually we emerged on the North Peak, in reality a knife-edged ridge on which are perched various temple buildings and a monastery. The ridge is so narrow that the track has to pass through the buildings, with no room on either side. From the ridge I gained my first full view to the plain below, the mountains of Shansi in the background and the great Yellow River flowing in between.

There were five Taoist priests on the North Peak and a boy who had been sent there from Shanghai by his parents for the benefit of his health. Here I was very kindly entertained as indeed I was everywhere on the mountain. Two days after my arrival Wolfram Eberhard, the distinguished German scholar, followed me up to the North Peak where he collected Taoist inscriptions and had long discussions with the priests. They were delighted to have with them someone who was so interested in their religion and had such a fine command of their language.

From the North Peak I went on to visit the other peaks of Hua Shan, the first being the West Peak, lying in a grove of pine trees and reached by a hair-raising track which in one place goes over what is known as the Sky Ladder. This was the end of the road for many pilgrims, including a famous T'ang personality called Han Yü. All around were spectacular pinnacles and rock walls, the scene continually changing through the interplay of sunlight and drifting mist. At the West Peak I met a priest whom I had known at the White Cloud Temple in Peking, and who took it upon himself to act as my guide and mentor from then on. Unlike me he had made the pilgrimage to Hua Shan on foot, having walked the whole way through the mountains of Shansi to Sian and then on to Hua Shan.

The South Peak is not really a peak at all but a gentle slope which was probably the area originally settled by hermits as it would have been possible to carry out a little cultivation there. The South Peak temple is the largest on the mountain, and the path from it to the East Peak is an easy one. Below the East Peak is a sheltered bowl in the mountain top. I was fortunate on the East Peak to witness a ceremonial Taoist dance carried out in slow time, said to represent the play of cosmic forces. One of the participants carried a fly whisk and the other a sword. The fly whisk, made from a yak's tail and known as a cloud sweeper, is thought to impart the ability to ride the clouds.

My days on Hua Shan passed all too quickly. I never met a more kind and sympathetic group of men than the Taoist priests on the mountain who seemed to derive real pleasure from the visit of Eberhard and myself. They even on one occasion dressed me up as a Taoist priest. The return journey was accomplished without incident though I found the descent of the Sky Ladder even more hair-raising than the ascent.

Later in the same year the mountain was visited by Miss Mullikin and Miss Hotchkis, the two indomitable and very gifted lady artists, American and Scot respectively, who travelled to all nine of the sacred mountains of China between 1935 and 1937. The account of their journeys, illustrated by their charming drawings and paintings, was published in Hong Kong in 1973.

The great West Peak of Hua Shan.

Refreshment shelter on the way up Hua Shan.

One of the sheer faces to be negotiated on the ascent.

View back to the North Peak from near the Sky Ladder.

Gnarled and contorted pine tree by the side of the track.

Artisan at work on an elaborate rendering of the character *shou* (longevity).

Ju-i sceptre, token of auspiciousness and good wishes.

Taoist monk reclining on an elaborately decorated bed.

Study of Taoist monk.

Study of Taoist monk.

Study of Taoist monk.

Study of Taoist monk.

Study of Taoist monk.

Study of Taoist monk.

The Lost Tribe Country 1936

IN 1936 I spent my holiday visiting the Lost Tribe country. This is an area of the further Western Hills about 160 kilometres from Peking lying up against a spur of the Great Wall in the vicinity of Ta Lung Men, the Great Dragon Gate. 'Lost Tribe' is not the translation of a Chinese term but one given it by foreigners. Despite its romantic title the Lost Tribe country is in reality a poor hill area inhabited by descendants of seventeenth-century rebels. These were part of a force led by a Shensi man called Li Tzu Ch'eng who in 1644 succeeded in capturing Peking while the main imperial forces were opposing the Manchus further north. When the rebels took the city the last Ming emperor committed suicide on Coal Hill.

Li's success was short-lived, however, for the Ming commander-in-chief, Wu San Kuei, then threw in his lot with the Manchus. Peking was quickly recaptured and Li's army dispersed. A group of his followers who had fled to the Western Hills later made their submission to the Manchus and were allowed to settle in the Lost Tribe country. They were not permitted to move and their descendants had been there ever since. I became interested in their history and in reports that the people still followed customs which had died out elsewhere.

Mr Bill Lewisohn, a British journalist and scholar who had travelled extensively in the Western Hills, helped me to make arrangements for the journey. For this I hired three donkeys and three donkeymen. Two of the donkeys were for carrying bedding and supplies, and the third was for me to ride on. But the riding donkey, black in colour, lacked the usual angelic temper of donkeys and early in the journey threw me off. I walked most of the way.

The donkeymen were very good people. They met me as arranged at the little railway station of Ch'ang Hsin Tien on the Peking-Hankow Railway and we set off. The route lay up the bare valley of the river known as Chü Ma Ho, and it took us several days to reach the Lost Tribe country. It was midsummer and extremely hot. By day we travelled through sparsely populated hills whose forest cover had long since been denuded, and at night we slept in temples which became smaller the further we went from Peking. We met few people on the way.

We had one alarm early on our journey. I had heard from the manager of the Hong-kong and Shanghai Bank, who had his information from the British Embassy, that there were bandits in the area. As we proceeded along a track in the bed of the valley several men came rushing down the hillside to meet us. Thoughts of banditry certainly

The three donkeymen who took me to the Lost Tribe country.

The head donkeyman cooking.

The rain god being carried through a village.

Decoration on a wayside shrine.

Shoemaker stitching on the sole, made of
cloth layers tightly sewn together.

Coiffures and ornaments in the Lost Tribe country.

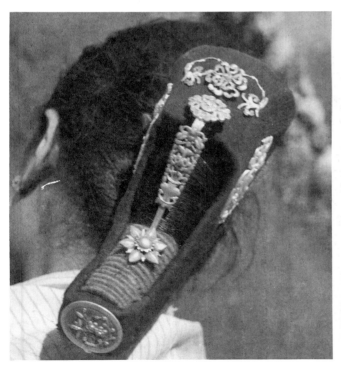

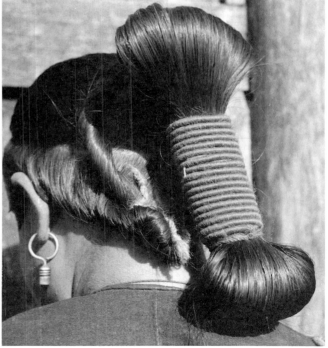

Young girl.

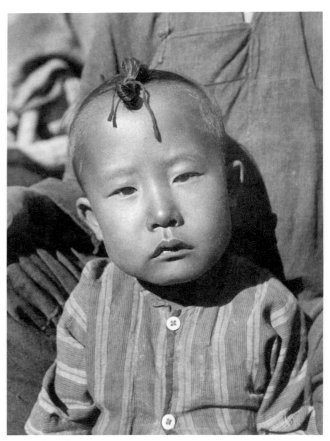

Small boy.

Lost Tribe girl.

View from the Great Wall towards the coastal plain at Shanhaikuan.

Net mending in Peitaiho, a small coastal village on the Hopei coast
which became a seaside resort for Europeans.

Junk under sail along the Shantung coast.

Junk under sail along the Shantung coast.

The master of the junk on which I travelled.

Pomfret for sale.

Cleaning fish for drying.

Purses embroidered with charming designs worn around the waist were
a feature of the attire of young girls in coastal villages.

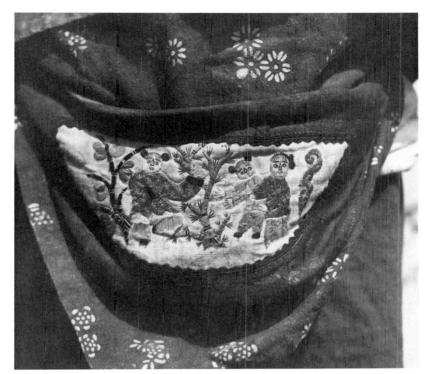

Embroidered designs cn purses.

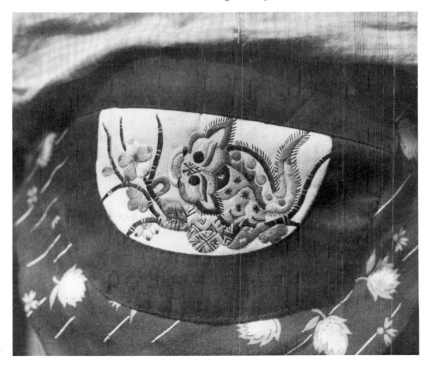

Houses in Shantung coastal villages were built of stone.

Even the poorer houses had massive stone walls.

Some brickwork was used for decorative purposes.
In this painted medallion four bats, symbols of happiness,
face the Eight Trigrams, which represent the universe.

Entrance to the courtyard of a well-to-do Shantung house.

Local celebration of the Dragon Boat Festival.

Two gaily clad girl acrobats.

A girl acrobat performing on an elevated stand, while the crowd's attention
is focused on a foreign woman with a camera.

The guardian of a local temple.

Village elder.

View over a coastal village.

Pao Ting 1940

AFTER Miss Bieber left Peking in 1940 I did pay one brief visit to the city of Pao Ting, about 150 kilometres to the south-west of Peking. Formerly the capital of the province of Hopei, Pao Ting had many fine relics. However, I was mainly interested in the street life. In Pao Ting I was the guest of Mr and Mrs Hugh Hubbard of the American Board of Foreign Missions. Mr Hubbard and his friend Dr George Wilder were pioneer ornithologists and the authors of a handbook on the birds of north-east China. Unlike some missionaries who were narrow-minded and bigoted, Dr Wilder and Hugh Hubbard represented the best kind of cultivated and open-minded mission worker. Dr Wilder and his wife, who had met at Oberlin College in Ohio in the 1890s, had been in China for more than forty years and had an excellent command of Chinese; they still studied Chinese texts regularly with a teacher. As far as they were concerned they were still learning.

A temple in Pao Ting.

Fruit stalls by rickshaw stand.

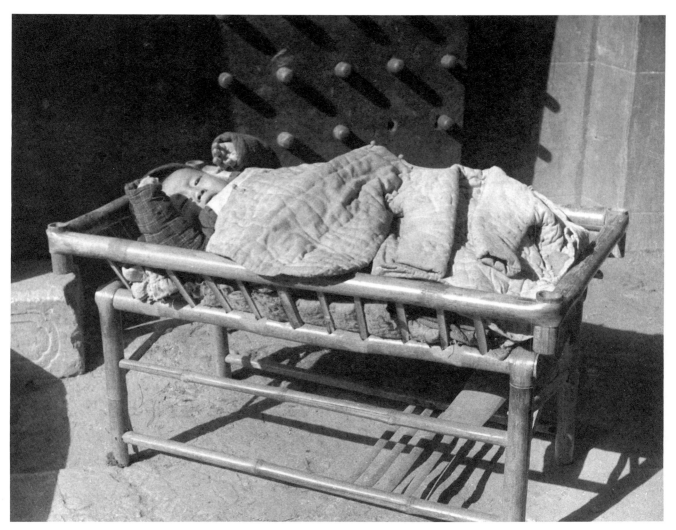

Baby's bamboo crib.

Ch'ü Fu and T'ai Shan 1942

IN 1942 I was asked by some German friends to go with them to T'ai Shan, a sacred mountain in Shantung, and to Ch'ü Fu, where Confucius lived and was buried. My role as guide for a group was not a satisfactory arrangement as it greatly restricted my ability to take photographs.

The area had been under Japanese occupation for several years and Japanese permits were needed. We started at Ch'ü Fu in the southern part of the province. Old Ch'ü Fu is several kilometres from the railway station because when the railway was built the descendants of Confucius refused to allow it any nearer. The old town was remarkable for its magnificent temple to the memory of the sage whose moral precepts guided China for some 2,500 years, and for the great cemetery where generations of his descendants are buried. Descendants were still living there, including the current holder of the title, Duke K'ung.

The trip was much too hurried and we did not attempt to visit the interesting but lesser known town of Tsou Hsien, some 28 kilometres south of Ch'ü Fu. Tsou Hsien was the birthplace of the sage Mencius who lived about 200 years after Confucius and who effectively revived and strengthened the teachings of the Master. It is almost as important a place as Ch'ü Fu in the history of Confucianism and is the site of another great temple and other relics. Here too descendants of Mencius were still living. I have always regretted not having been able to visit the Ch'ü Fu-Tsou Hsien area at leisure and in my own time.

From Ch'ü Fu we went north to the town of T'ai An which lies below the sacred mountain of T'ai Shan. Rising to a height of 1,545 metres, T'ai Shan is a bare stony mountain held in great veneration by Taoists. Although not nearly as spectacular as Hua Shan, nor its temples so well cared for, T'ai Shan was still being visited by large numbers of pilgrims. The ascent was arduous: from T'ai An some 5,900 steps had to be mounted to reach the top. The well-to-do were generally carried up in sedan-chairs, the only alternative at the time to walking. Today an aerial sedan-chair (cable car) carries the visitor most of the way up the mountain. Like so many facets of life in the old China the sedan-chair business was highly organized: a special guild had the monopoly and for some curious reason all the guild members were Muslims.

My friends and I made the ascent on foot which was very exhausting in the summer heat. The old road is a paved one running up a valley on the flanks of T'ai Shan. Originally the valley was clothed with groves of cypress and pine trees, but many had been felled. About half way up there was an important temple, the Middle Gate of Heaven, which, after a fairly level stretch, led to the great stairway of some 2,000 steps to the summit. In clear weather the site commanded a spectacular view over southern Shantung. On the summit were a number of fine temples and we spent the night in one of these. I was so busy looking after my companions that I kept no notes, and unfortunately my recollections of the trip are dim.

After T'ai Shan we visited one unusual temple reached from a stop on the railway between T'ai An and Chi Nan. This was Ling Yen Ssu. It had some fine old buildings, including a pagoda, and was notable for some life-size figures clad in the Chinese style but with European facial characteristics. I have never found a full description of the temple in the works I have been able to consult.

I accompanied my friends to Tsingtao where I had the misfortune of coming down with scarlet fever. I had been feeling unwell during much of the trip. There was a well-established German hospital in Tsingtao but it refused to admit me. A good Samaritan, a Mrs Boetcher who ran a small guesthouse, quietly took me in and kindly took care of me without alarming the other guests. I was treated by a German missionary doctor, Dr Eitel, a good and liberal man. With their help I made a quick recovery and was able to return to Peking. Dr Eitel was later recalled to Germany on one of the blockade-breakers, merchant navy ships which essayed the journey back to Europe. Some of these blockade-runners got through but I heard that Dr Eitel's ship was not so lucky and that he was lost on the journey.

The Temple of Great Harmony, commonly known as the
Temple of Confucius. The elaborately carved dragon pillars
were erected in the year 1500.

Statue of Confucius.

The memorial arch on the spirit road to the funeral grove.

Grave of the son of Confucius.

The intensely cultivated farmland around Ch'ü Fu.

Loads were suspended from the ends of a long pole
carried on the shoulder.

Looking down the T'ai Shan track.

Looking up the T'ai Shan track.

The last flight of steps to the summit of T'ai Shan.

The T'ai Shan Goddess, to whom, in former days, many valuable offerings
were made, to the great benefit of temple revenues.

Temple scene, T'ai Shan.

Ling Yen Ssu was noted for 40 life-like figures
of Lohan, disciples of the Buddha.

The figures date back to Sung times.

Street scene in Chi Nan, the junction for the Tsingtao railway.

Street scene in Chi Nan.

Street scene in Chi Nan.

Nanking 1944

MY last venture outside Peking was in 1944. The German Ambassador to the puppet government of China, Dr Woermann, invited me to compile a photographic record of Nanking. Dr Woermann was an unconventional bachelor diplomat more interested in art than politics. Funds were available to him for cultural activities and he decided to make use of them to produce a book on the city, for which I was to take the photographs and the German scholar, Alfred Hoffmann, was to provide the text.

As a result I spent some time in Nanking in the summer of 1944. Nanking has an ancient history, but few cities in the world have known so much destruction and pillage over the centuries. Its importance stems from its geographical location, the city lying against hills on the south bank of the Yangtze at a point where that great river is confined to a channel no more than 1,100 metres in width. The area around Nanking is rich and fertile, and in imperial times the city was always the main administrative and trading centre of the populous lower Yangtze valley.

Nanking's importance is demonstrated by the way in which it has repeatedly recovered from terrible devastation. Twice in its history it had been destroyed and at other times it had known other calamities. The most recent of these was the savage sack of the city and large-scale massacre of its inhabitants by the Japanese after their capture of the city in December 1937. Yet Nanking has always recovered. When I was there in 1944 it again had a considerable population despite the savagery of the Japanese only seven years earlier.

There are many prehistoric remains around Nanking but its recorded history commences from about 500 BC, at the beginning of the Warring States period, several centuries before China became a unified empire in the late third century BC. This unity lasted until the early third century AD and was followed by the long and complicated history of rivalry between the Chinese dynasties of central China and the largely non-Chinese rulers of the north. For much of the period between 229 BC and AD 589 Nanking was the capital of south China although the main political capital remained in the north. Its wealth and power generated hostility in the north and upon its capture by the short-lived Sui dynasty in AD 589 it was almost totally destroyed.

In subsequent years Nanking steadily regained its importance as a regional and cultural centre but decline again set in during the Yüan dynasty under the Mongols. It acquired its greatest glory during the early years of the Ming dynasty, whose founder came from a family which originated in the neighbouring province of Anhui. Nanking was the imperial capital from 1368 to 1420, in which year Emperor Yung Lo moved the capital to Peking, retaining Nanking as the second capital. It continued to serve as the main regional centre of central China and as a centre of art and scholarship, a situation which remained unchanged for much of the succeeding Manchu dynasty.

A notable event in the history of Christianity in China occurred in Nanking, for it was here in 1707 that the Papal Legate published the edict forbidding Chinese Catholics to practise rites honouring Confucius and their ancestors. These rites had been acceptable to the famous Jesuit missionaries who had brought Christianity to China in the sixteenth century and who had acquired much influence at court and in official circles. The edict greatly reduced the spread of Christianity in China.

The decline in Manchu power and authority in the nineteenth century was reflected in the fortunes of Nanking. The Treaty of Nanking in 1842 ending the first Opium War and signed under duress on a British warship was the first of the many setbacks and humiliations that China had to endure at the hands of Western powers. Only a few years later Nanking became the storm centre of the disastrous T'ai P'ing rebellion.

Historically China has known many rebellions but the T'ai P'ing episode was a rebellion with a difference. It was led by a failed Cantonese scholar who turned to Christian teachings and who came to believe that he had a heavenly mission to reform China. These beliefs gave the movement an ideological base previously unknown in China. Central China was quickly overrun by the T'ai P'ing and only disunity in its leadership prevented the movement from taking north China and capturing Peking. The leader adopted the title of Heavenly King and the capital was established in Nanking.

In many ways the T'ai P'ing rebellion was a genuinely reformist movement whose adherents believed in the Christian God and were guided by the Ten Commandments. But as power brought corruption, the movement lost its early drive and efficiency and it was weakened by internal dissension. The foreign powers initially maintained neutrality and after the conclusion of the second Opium War gave the imperial authorities direct and indirect aid in suppressing the movement. This was achieved in 1863.

Unfortunately the T'ai P'ing rebellion and its suppression caused immense cultural and human loss; many traditional places of worship and monuments were destroyed and the toll of life was enormous. Nanking went into decline, and although its population grew again it was not to recover as a political centre until the overthrow of the Manchus.

It was the intention of the revolutionary movement led by Dr Sun Yat Sen to establish the new capital of a republican China in Nanking. A declaration to this effect was made in the city on 1 January 1912. Politicians in the conservative north, however, wanted a constitutional monarchy. Eventually a compromise was reached whereby the republican constitution was accepted, the northerner Yüan Shih K'ai becoming president and the capital remaining in Peking.

Peking remained the capital until 1927 when the Kuomintang under Chiang Kai Shek finally established Nanking as the seat of government. Captured and brutally ravaged by the Japanese in 1937, Nanking became the capital of a puppet regime led by Wang Ching Wei, one of the early anti-Manchu revolutionaries and a rival of Chiang Kai Shek. Wang Ching Wei died in 1944. The People's Republic re-established the national capital in Peking, a reversion to the traditional location of the capital close to the northern frontier.

The result of Nanking's eventful and often violent history is that the innumerable historical remains of great interest and antiquity in and around the city are largely ruins. Some of the buildings have been restored but few are originals. Nevertheless many relics of the past were to be seen when I visited Nanking, and more are being located. Of particular interest has been the discovery in 1950 of the tombs of the first two emperors of the Southern T'ang dynasty (AD 937–75) and the identification in 1958 of the tomb of a visiting king of Borneo, an ancestor of the present Sultan of Brunei, who died in Nanking in 1408.

In 1944 the most conspicuous relics were those dating from the Ming period. The Ming city wall stretched for 38 kilometres, much of it still in place though in a ruinous state. As can be seen from the map (see map at the end of the book) its shape is irregular. The wall is built of brick on stone foundations and its dimensions vary from place to place. The bricks carry seal impressions indicating their provenance, and from them, historians have been able to establish that the bricks were supplied from many districts, mainly in the provinces of Kiangsu, Anhui, Kiangsi and Hupei.

The Ming Drum Tower, which had been well maintained and restored, was especially prominent. Built on a slight eminence, the tower was where drums were beaten to announce the changes of the night watches. A little way to the north-east a small pavilion, built in 1889, contains a great bell which was cast in Ming times to complement the Drum Tower. Another interesting remnant of the Ming period existed in the southern part of the city: a tower which used to stand in the centre of the rows of hundreds of cells to which students were confined while they sought to obtain the coveted qualification opening the door to a career in the civil service. The cells are gone but the tower remains, unlike the situation in Peking where no trace of the examination halls has survived. On the other hand little remains of the splendid Ming palace which must have resembled the Forbidden City in Peking. It was allowed to decay under the Manchus.

More could be seen of the mausoleum of the first Ming emperor, Hung Wu, which is situated in the hills known as the Purple Mountain to the east of the city. It has lines of stone animals and officials similar to those of the Ming Tombs near Peking, but the site is much less spacious and the figures more closely grouped together. The entrance-way along which the animals are grouped runs first westward, then northward and finally eastward, so as to prevent the entry of malign influences which were believed to travel always in straight lines.

Originally there was a temple here, but it was demolished to make way for the mausoleum and re-erected a little distance away to the east. This is Wu Liang Tien or the Beamless Hall, the only Ming building in Nanking that remains intact. It was used by the Kuomintang as a memorial to fallen soldiers.

There are many other Ming remains in and around the city, principally the tombs of important early Ming officials. An unusual and interesting Ming relic is the limestone quarry which provided much of the raw material for buildings and monuments. The quarry is notable for one enormous monolith commissioned by Emperor Yung Lo for erection at his father's tomb. It was realized only at a late stage that even Chinese technical ingenuity was incapable of moving it and the monolith has remained in the quarry ever since.

Less conspicuous but still numerous are relics of the dynasties that preceded the Ming. Some of the most important, consisting of tombs, date back to the Liang dynasty (AD 502–57). The tombs of the Liang emperors themselves are not in Nanking, but are located at the ancestral home of the Liang family at Tan Yang, 70 kilometres east of Nanking, which I did not visit. North-east of Nanking are several

tombs of members of the Liang royal family. They contain some fine statues, especially impressive being those of open-mouthed winged lions.

In the same area is Ch'i Hsia Ssu, a monastery which also dates to the sixth century.
Originally the haunt of hermits, it became under the T'ang dynasty (AD 618–907) one
of the four greatest monasteries of China. In caves near the temple are many heavily
restored Buddhist statues. Ch'i Hsia Ssu was almost totally destroyed by the T'ai P'ing
rebels, the present buildings having been erected after Nanking was recaptured. It
remains, however, a very beautiful relic. Adjoining Ch'i Hsia Ssu is the small She Li
Pagoda. It is one of the two best preserved T'ang buildings in Nanking, the other
being the handsome reddish-coloured pagoda situated south of the city on Niu T'ou
Shan.

There were many other reconstructed temples in Nanking, some of them of great
charm. The most important was the Temple of Confucius in the western part of the
city. In the centre of Nanking, to the east of the Bell Pavilion, was the Cock Crow
Temple (Chi Ming Ssu). The site was a Mongol execution ground and the original
temple was built to placate the ghosts of those killed there. P'i Lu Ssu, also in central
Nanking, contained many small images of the Buddha. On a hill overlooking the
Yangtze downstream from Nanking was a very picturesque temple known as the
Swallow's Rock Temple. There were many more.

After the re-establishment of Nanking as the capital in 1927 the Kuomintang
undertook considerable development of the city, constructing roads and public buildings. The most famous of the buildings from this period is the mausoleum of Dr Sun
Yat Sen, built on the Purple Mountain a little to the west of the mausoleum of Hung
Wu. It is an impressive structure which was largely financed by subscriptions from
overseas Chinese. From a white marble entrance *p'ai-lou* a flight of 392 steps leads up
to the mausoleum; both *p'ai-lou* and mausoleum are roofed with blue tiles. North-west
of the Sun Yat Sen mausoleum was a modern observatory where some ancient
instruments were exhibited. A curiosity of the observatory hill was an imposing
modern pagoda designed in the 1930s by an American architect named Murphy.

In ancient times Nanking was noted for its gardens. Like so many facets of Chinese
culture, gardens developed in an entirely different way to those of the West. Although
flowering plants did play a role in Chinese gardens, much of the emphasis was on
rockeries, pieces of eroded limestone pitted with many cavities of different shapes and

A corner of the Ming city wall in the northern part of Nanking.

View from a section of the city wall.

All that remains of one of the entrance gateways to the Ming Palace.

Stone animals flanking the approach road to the mausoleum of
the Ming emperor Hung Wu on the slopes of the Purple Mountain.

The Beamless Hall adjacent to the emperor's tomb
built to replace the temple which originally
occupied the site of the mausoleum.

The Ming quarry which provided the stone blocks
for statues and memorial tablets.

The Nanking mosque.

Interior of the Nanking mosque.

Courtyard of Ling Ku Ssu, a Buddhist temple north of Nanking.
A neighbouring pagoda, dating to the 1930s, was the work of
an American architect named Murphy.

The garden of P'u Te Ssu.

The Swallow's Rock Temple overlooking the Yangtze
north of Nanking.

View over the Yangtze from the Swallow's Rock.

The theme of 'Phoenix Paying Homage to the Rising Sun'
as depicted on a Ming ornamental carving.

Stone figures at the tomb of a Ming official.

Low pillar surmounted by a lion playing with a brocaded ball.

Vignette in a Nanking street.

A modest funeral in Nanking.

The approaches to the Sun Yat Sen Mausoleum.

Nanking University.

Street scene.

A modern building in the centre of Nanking.

Street scene.

Outer walls of a rich man's residence.

Approach to a rich man's residence.

View from the Temple of Confucius looking over the city
towards the Purple Mountain.

Early morning on the Lotus Lake, with the Purple Mountain
in the background.

Scoop-net fishing, a common sight along
the streams and canals.

Collecting reeds for mat making.

Elderly peasant.

Depilating unwanted hair to accentuate the extent
of the forehead.

The umbrella maker.

Market woman at work.

A cooper at work.

Warming up in the morning sun after a night of low temperatures.

Laying out dried noodles before packing.

Shop selling rope and string.

Weaving coarse cotton cloth.

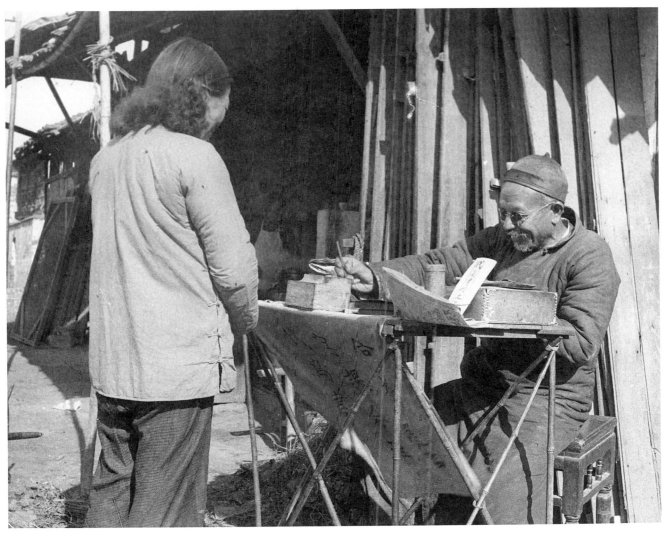

Letter writer at work.

Fortune-teller.

An ingenious fortune-teller who could use several brushes simultaneously.

Bibliography

General

The Cambridge Encyclopaedia of China (1982).
Nagel's Encyclopaedia-Guide, China (1979).
Guide to China (Japan Government Railways, 1924).
The Encyclopaedia Sinica (Shanghai, 1917; reprinted by Oxford University Press, 1983).

Jehol

Sven Hedin, *Jehol, City of Emperors* (New York, 1933).

Hua Shan

Hedda Morrison and Wolfram Eberhard, *Hua Shan* (Vetch and Lee, Hong Kong, 1974). (The book was commissioned by Henri Vetch of the firm of Vetch and Lee of Hong Kong. Vetch was formerly proprietor of the Peking Book Store and he settled in Hong Kong after his imprisonment and release by the Chinese Communists. He was intensely interested in Taoism, and he lavished the most scrupulous attention on every aspect of the production of the book. Sadly the photographs were printed on inferior paper.)

Hua Shan and T'ai Shan

Anna M. Hotchkis and Mary A. Mullikin, *The Nine Sacred Mountains of China* (Vetch and Lee, Hong Kong, 1973).

T'ai Shan and Ch'ü Fu

Father F. Dransmann, *T'ai Shan–Kufow Guide* (The Catholic Mission Press, Yenchowfu, 1934). (Father Dransmann was a missionary of the Society of the Divine Word. The book is notable for its excellent photography and it is to be hoped that his photographic collection has survived.)

Nanking

Father Louis Gaillard, SJ, *Nankin d'alors et d'aujourd'hui* (*Variétés Sinologiques*, No. 23, Catholic Mission Press, Shanghai, 1903). (This is the most detailed and scholarly Western-language work on the city.)

Alfred Hoffmann and Hedda Hammer, *Nanking* (1945).

Barry Till and Paula Swart, *In Search of Old Nanking* (Hong Kong, 1982). (Though not really comparable to its namesake, *In Search of Old Peking*, this is a very useful little book containing a great deal of information. Strong on tombs, weak on temples.)

Romanization

THE following is a list of the Chinese names and terms as they are romanized in this book, with their equivalents in the Pinyin system of romanization currently used in the People's Republic of China, as well as the Chinese characters for them.

Anhui	安徽	Anhui
Ch'ang Hsin Tien	長辛店	Changxindian
Ch'eng Hua	成化	Chenghua
Ch'eng Te	承德	Chengde
Cheng Ting	正定	Zhengding
Ch'i Hsia Ssu	棲霞寺	Qixiasi
Chi Ming Ssu	雞鳴寺	Jimingsi
Chi Nan	濟南	Ji'nan
Chia Ch'ing	嘉慶	Jiaqing
Chiang Kai Shek	蔣介石	unchanged
Ch'in Huai	秦淮	Qinhuai
Ch'ing	清	Qing
Ch'ü Fu	曲阜	Qufu
Chü Ma Ho	拒馬河	Juma He
Chungking	重慶	Chongqing
Fukien	福建	Fujian
Han Yü	韓愈	Han Yu
Hopei	河北	Hebei
Hsien Feng	咸豐	Xianfeng
Hsü Mi Fu Shou Chih Miao	須彌福壽之廟	Xumifushouzhimiao
Hsüan Wu	玄武	Xuanwu
Hua Shan	華山	Huashan
Hua Yin	華陰	Huayin
Hung Wu	洪武	Hongwu
Hupei	湖北	Hubei
Jehol (Je Ho)	熱河	Rehe
ju-i	如意	*ruyi*
k'ang	坑	*kang*

K'ang Hsi	康熙	Kangxi
Kiangsi	江西	Jiangxi
Kiangsu	江蘇	Jiangsu
K'ung	孔	Kong
Kuomintang	國民黨	Guomindang
Kwangtung	廣東	Guangdong
Li Tzu Ch'eng	李自成	Li Zicheng
Liang	梁	Liang
Ling Ku Ssu	靈谷寺	Linggusi
Ling Yen Ssu	靈巖寺	Lingyansi
Lohan	羅漢	Luohan
Lu Kou Ch'iao	蘆溝橋	Lugouqiao
Lung Hsing Ssu	隆興寺	Longxingsi
Mencius	孟子	Mengzi
Ming	明	Ming
Nanking	南京	Nanjing
Niu T'ou Shan	牛頭山	Niutoushan
p'ai-lou	牌樓	*pailou*
Pao En Ssu	報恩寺	Bao'ensi
Pao Ting	保定	Baoding
Pei Hua Shan	北華山	Beihuashan
Peitaiho	北戴河	Beidaihe
Peking	北京	Beijing
Peking-Hankow Railway	京漢鐵路	Beijing-Hankou Railway
P'i Chih Pagoda	辟支塔	Pizhi Pagoda
P'i Lu Ssu	毗盧寺	Pilusi
P'u Lo Ssu	普樂寺	Pulesi
P'u Ning Ssu	普寧寺	Puningsi
P'u Te Ssu	普德寺	Pudesi
Shanhaikuan	山海關	Shanhaiguan
Shansi	山西	Shanxi
Shantung	山東	Shandong
She Li	舍利	Sheli

Sheng Mu	聖母	Shengmu
Shensi	陝西	Shaanxi
Shih Chia Chuang	石家莊	Shijiazhuang
shou	壽	*shou*
Sian	西安	Xi'an
Sun Yat Sen	孫逸仙	unchanged
Sung	宋	Song
Ta Fo Ssu	大佛寺	Dafosi
Ta Lung Men	大龍門	Dalongmen
Ta T'ung	大同	Datong
T'ai An	泰安	Tai'an
T'ai Miao	泰廟	Taimiao
T'ai P'ing	太平	Taiping
T'ai Shan	泰山	Taishan
T'ai Yüan	太原	Taiyuan
Tan Yang	丹陽	Danyang
T'ang	唐	Tang
Taoism	道教	Daoism
Tsin Ling	秦嶺	Qinling
Tsingtao	青島	Qingdao
Tsou Hsien	鄒縣	Zouxian
Wang Ching Wei	汪精衛	Wang Jingwei
Wei	魏	Wei
Wei Hai Wei	威海衛	Weihaiwei
Weng	翁	Weng
Wu Liang Tien	無樑殿	Wuliangdian
Wu San Kuei	吳三桂	Wu Sangui
Yangtze	揚子	Yangzi
Yü Hua T'ai	雨花臺	Yuhuatai
Yüan	元	Yuan
Yüan Shih K'ai	袁世凱	Yuan Shikai
Yün Kang	雲岡	Yungang
Yung Lo	永樂	Yongle

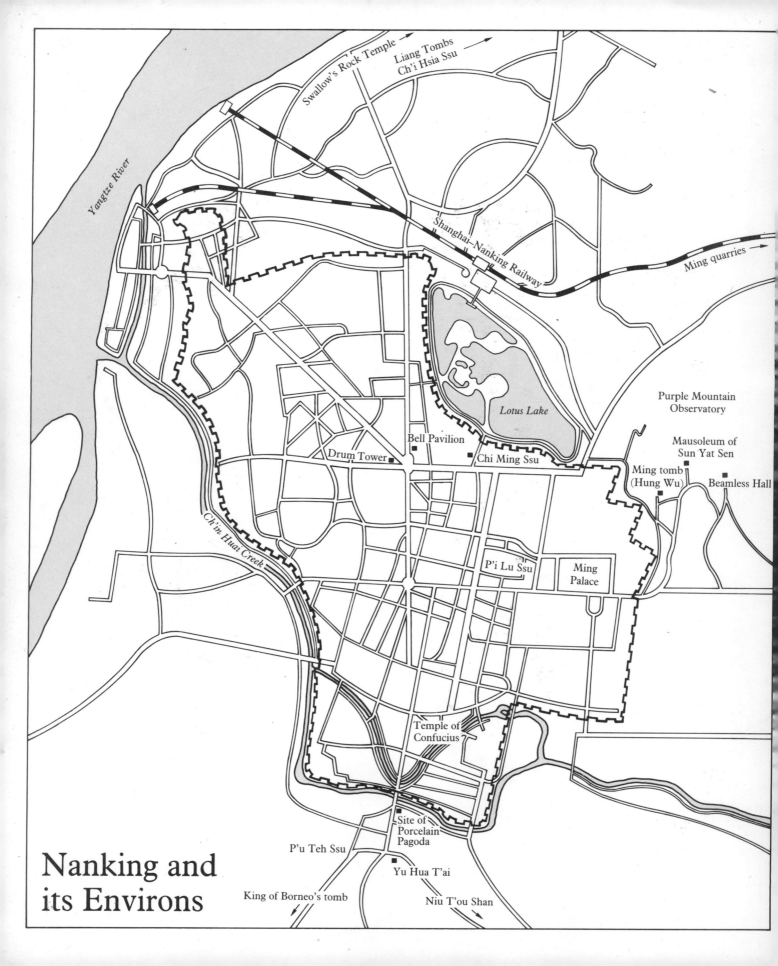

Swallow's Rock Temple

Liang Tombs
Ch'i Hsia Ssu

Yangtze River

Shanghai–Nanking Railway

Ming quarries

Lotus Lake

Purple Mountain
Observatory

Bell Pavilion

Mausoleum of
Sun Yat Sen

Drum Tower

Chi Ming Ssu

Ming tomb
(Hung Wu)

Beamless Hall

Ch'in Huai Creek

P'i Lu Ssu

Ming
Palace

Temple of
Confucius

Site of
Porcelain
Pagoda

P'u Teh Ssu

Yu Hua T'ai

King of Borneo's tomb

Niu T'ou Shan

Nanking and
its Environs